Out in the Night

For children all over the world, the night is aglow with wonder. They peer out their windows, out their doors, hoping to see what creature caused that rustle . . . that cry . . . that pungent smell in the darkness around them.

Shhh! Where is that strange howl coming from? That grunt? That snort? Did you see that shadow gliding through the bushes over there? Beneath the starry sky it is time to squint your eyes, perk up your ears, and discover the nocturnal animals sharing our Earth . . . out in the night.

Harbinger House, Inc.
Tucson

Art Director, Ellen McMahon

 Published by Harbinger House, Inc., 3131 N. Country Club, Tucson, Arizona 85716.

Library of Congress Cataloging in Publication Data on last page.

Printed in Singapore by Singapore National Printers, Ltd. through

Out in the Night

by Karen Liptak
illustrated by Sandy Ferguson Fuller

HARBINGER HOUSE JUVENILE NATURAL HISTORY SERIES

IN THE TROPICS OF SOUTHEAST ASIA

In the shadow of the children's house in Malaysia, the unusual scaled anteater known as a pangolin hunts for ants tonight. But the children do not see him. They're too busy staring at a tree that has turned into a festival of light tonight! The whole tree twinkles as thousands of fireflies flash their eerie lights. Fireflies aren't really flies at all; they're soft-bodied beetles. Each species of this unusual insect has its own unique mating flash.

In another tree, a green and black flying snake leaps to within inches of a furry, large-eyed colugo (also known as a flying lemur) foraging upside down for leaves and flowers. But neither creature can actually fly. Instead, they make gliding leaps among the trees, the natural home of many unique animals in this tropical climate.

Near their porch, the children's affectionate bush-tailed binturong snacks on some tender young plants. While it is illegal to keep wild animals as pets in Malaysia, many animals, such as this catlike mammal, live in the villages. They are never caged.

As usual, other fascinating creatures are out on this hot midsummer's night. A lesser mouse deer, actually related more to camels than deer, disappears into the undergrowth with a dainty gait. An alert muntjac, his large ears upright, chews on the bark of a tree. This little deer is a herbivore, a plant-eating animal. Near him, a Malaysian porcupine shuffles along, searching for fallen fruit. And that smell like rotten garlic? It could mean that a moonrat is near. This solitary black and white mammal with a long snout belongs to the hedgehog family, and primarily eats insects.

Overhead, a small Malaysian eagle owl silently soars through the sky. Will she snatch that foot-long walking stick stretching out on a branch now that it's dark? Or will the walking stick fool the owl into thinking he's just another twig?

The children recognize that honking sound. The beautiful raspberry-colored Asiatic painted frogs that live in their mother's empty flowerpots during the day are calling to their mates.

And the loud, barking "geek-oh?" The children smile, since the noise comes from a house-dwelling gecko known as a tokay. It is considered good luck to hear the voice of this orange and red lizard indoors, even if he wakes you up at night. But keep your fingers away from his mouth. Tokays bite!

ALONG THE OREGON COAST

What exquisite sea animals the low tide has exposed in the tidal pool for the children to discover.

Amidst the twinkling reflections of the stars overhead a lovely green sea anemone waves his delicate tentacles in the water — tentacles that will poison any small fish that brushes up against them. And there's a beautiful vermillion starfish, which can regrow any of her five arms should they break off. Giant red sea urchins are in the tidal pool as well. Close cousins of the starfish, they are called "porcupines of the sea" because of their sharp, protective spines.

Beside the tidal pool, a ten-legged hermit crab is busy househunting tonight. The crab taps an empty seashell with its claws and antennae. Will this make a good home? The hermit crab must move each time he outgrows the house he is in.

Meanwhile, two purple shore crabs have come face to face in their scavenging for food. They threaten each other with their pincers open wide, both ready to do battle until one loses a claw and scuttles away. Like their colorful neighbors in the tidal pool, crabs are *invertebrates*, animals without a backbone.

Other invertebrates, known as limpets, slowly crawl over the rocks, seeking algae for supper. Limpets are small sea animals that live in a protective shell shaped like an upside-down bowl.

Familiar sounds pierce the salty night air: the "kee-a, kee-a" of a flock of white western gulls disturbed from sleep. Children on the beach can also hear the occasional barks of playful stellar sea lions, the largest of all sea lions. These endangered marine mammals live in colonies all along our Pacific coast. They are protected by law and can no longer be hunted.

What was that strange "whoosh?" The children look out across the waves just in time to see the spray of two huge grey whales heading north from the Gulf of Mexico to their summer feeding grounds in Alaska. Like the sea lions, whales are *mammals*, warm-blooded animals who produce milk for their young. It's a special thrill to see the whales so close to shore.

ALONG THE AMAZON IN BRAZIL

Near the old city of Manteus, on the Amazon River, the jungle is waking up. The children linger on their porch, reluctant to go to bed.

Their pet parrot is quiet now, but the warm night is full of sound. Is that a woodpecker they hear? No, it is a spectacled owl, called a "knocking owl" because of her rapping call. And that soft, scuffling noise? It is coming from a coarse-haired lesser anteater at the base of the tree. He is ripping open a termite nest with his powerful claws. The toothless anteater will scoop up as many as 500 of these insects at once with his long, sticky tongue.

Higher up in the tree, bony-headed tree frogs leap into the air, catching insects. The disks on their toes help these graceful amphibians stick to the trees. Like all amphibians, frogs are *vertebrates*, animals with a backbone. Fishes, reptiles, birds, and mammals are also vertebrates.

The frogs must beware. A cat-eyed snake is on the prowl, with big, vertical pupils well suited for night vision. And frogs are a favorite prey. Another snake, a beautiful emerald tree boa, hangs from the tree, mouth open, waiting for a careless bird to fly within his striking range.

High overhead, the branches rustle and sway beneath the weight of squeaking night monkeys traveling the "highways" they use among the trees. Sometimes the children can catch a fleeting glimpse of these acrobats whose huge brown eyes seem to pop out of their white faces. But not tonight. Nor do they see the giant armadillo waddling to the river's edge for a drink, or the fish-eating bat raking the water's surface with his claws.

For a few moments the children watch the piper-eating bat dine on a small, thick-stemmed piper plant. Many of this bat's jungle neighbors also depend upon the piper plant for food. The bat eats the plant's tiny green fruit and then disperses its seeds, making her one of the Amazon's most important residents.

Was that a human cough? Or could it be a jaguar, the Amazon's largest and most secretive wild cat? The jaguar is a *carnivore*, a flesh-eating animal. Luckily, he sounds far away.

And, oh no, not again! The children's pet tapir has broken his leash and is feasting on their mother's favorite orchids. Because of his destructive ways, the children will soon have to release the big-snouted baby back into the jungle.

June is a winter month in Australia, home of many animals that are found nowhere else in the world. Ranchers consider many of them pests.

But children disagree. They love to watch the "mobs" of red kangaroos leaping across the scrub-lands at speeds of up to forty miles an hour. The 200-pound reds are the largest of Australia's forty species of "roos." All kangaroos are *marsupials*, mammals that carry their young in pouches. A "joey," or young kangaroo, lives in his mother's pouch until he is at least eight months old.

Many other Australian mammals are also marsupial night-roamers. The tree-dwelling brush-tailed wambenger, looking for a bird's nest to rob tonight, is a marsupial mouse.

Another marsupial mouse, the fat-tailed dunnart, leaps up to catch a moth. He can eat more than his own weight each night. In times of plenty he stores food in his tail. The tail will actually look much like a tiny carrot. Then when food is scarce the tail provides the fat needed for the dunnart to survive.

Overhead, two playful brush-tailed possums greet each other, their ears erect, before proceeding on their way. These furry marsupials eat insects, birds' eggs, leaves, and fruit — and are not beyond raiding garbage cans for their dinner.

Behind them stands a huge termite mound, a marvelous example of the termites' architectural skill. Inside the mound, these sightless insects are constantly busy making repairs, gathering food, and building tunnels. Termites have a complex social system made up of soldiers and workers, all organized under one "queen mother."

Reptiles are also out tonight. Most of Australia's snakes are poisonous, although the colorful bandy bandy on the eucalyptus tree is harmless to humans. But the smooth knob-tailed gecko flicking up her tongue to clean her lidless eyes had best look out! Like the dunnart, the gecko has a tail that stores fat for leaner times. Like many other geckos, she can shed her tail if trapped. A new one will grow later.

"Kook kook." A spotted nightjar cries out her threat to protect her young, which are camouflaged in their leaf litter nest on the ground. But the nightjar's call is overpowered by the "Morepork, morepork" of the small boobook owl.

The boobook's call is as much a part of Australia's night as the carolling of magpies is a part of its day. Now that the boobook's mating season has begun, children will hear him serenade his lady love until dawn. "morepork, morepork, morepork."

UP IN THE ARCTIC

It is July, time of the "midnight sun" in Samiland (also called Lapland). During the Arctic's two months of summer, the unsinking sun shines twenty-four hours a day. The whole pattern of life on the tundra changes as the animals make the best use of this constant sunlight.

Close to the fence around the children's house the family's reindeer rest in the pasture, their work done for the day.

Outside the fence, an Arctic fox, camouflaged in his summer coat of brown, stalks the favorite food of many Arctic animals. These are rodents known as lemmings. The number of lemmings changes from year to year, depending on the number of plants for them to eat. In turn, the animal population that feeds upon lemmings fluctuates too. This summer the lemmings are plentiful, so the wildlife in Samiland is abundant.

Another camouflaged mammal, an Arctic hare, scurries away. The round-the-clock sunlight makes nocturnal activity possible for the hare, as well as for many other animals that are usually *diurnal*, or active by day.

The wheatear bird is one more Arctic resident normally asleep at night. But for the next few weeks the mother bird will constantly feed her newly-hatched young hidden in a nearby crevice. Her chicks must be able to fly before the short summer is over. The lantern-eyed snowy owl is also a busy parent, mindful of protecting her young from predators. The owl's nest is a mound shaped like an empty volcano, built on high rocky ground.

While the snowy owl has little company during the winter, many birds share her environment now that summer is here. Thousands of these migrants fly north as soon as the ice cracks on the tundra's lakes. They feed upon the millions of swarming mosquitos and other biting flies that also make the summer tundra their home.

A sweet "dip, dip, dip" sound catches the attention of the children inside. It comes from a male bluethroat, a close relative of the nightingale, that is warbling away right on the fence!

And what is that? The howls of Arctic wolves harmonize in a haunting refrain. But it is rare for children to see these sharp-eyed hunters, for the Arctic wolf is nearly extinct.

IN THE SAHARA DESERT

It is a lovely June evening in Algeria. Outside the thousand-year-old city of Ghardaia, a Berber boy sits with his grandfather, watching the moon rise. The family and their camels have traveled far today. The camels close their weary eyelids, three on each eye to help keep out sand and dust.

But the boy eagerly listens to the sounds of the night. A shriek! Is it a dog within Ghardaia's stone walls? Or a lone jackal many kilometers away?

Closer by, a tiny fennec fox softly rustles about as he hunts for food. The fox's enormous ears give him keen hearing. They also help him lose excess body heat and keep comfortable in this hot, dry climate.

Many more inhabitants of the Sahara have also left their daytime homes. Ages ago, their ancestors discovered that under the cover of darkness they had less competition for food and more concealment from enemies. Then, too, the desert's cool nights were more tolerable than its scorching days. Today, especially in desert and tropical environments, more animal activity goes on by night than by day.

A pipistrelle bat prowls the sky for insects to eat. Like most bats, the little pipistrelle navigates by a process called *echolocation*. The bat makes high-pitched squeaks, then interprets the echoes when the sounds hit something.

A large eagle owl is also in the sky. The owl's sharp vision and hearing make him a skilled predator in the dark. His fluffy wings help him surprise his prey in silence. Spotting the eagle owl, a long-tailed jerboa hops away like a tiny kangaroo to the safety of her underground tunnel. The jerboa is actually a rodent, as is her neighbor, the gerbil, or "sand rat." He pounds his long tail on the ground to warn other sand rats of the danger above. Danger lurks on the ground, as well.

A poisonous horned viper's eyes are glued on the gerbil. This sidewinding snake has wiggled his body into the sand so well that a small web-footed gecko steps behind him with no idea his deadly enemy is just millimeters away.

But the gecko, a nocturnal lizard, is careful not to bump into the two scorpions "holding claws" in their strange form of courtship. Scorpions are night crawlers that belong to the same class of animals as spiders.

Although the Berber boy can't see the scorpions, he can see a flock of European swallows across the full moon as they migrate from South Africa to Europe for the summer. A wondrous sight in the Sahara night!

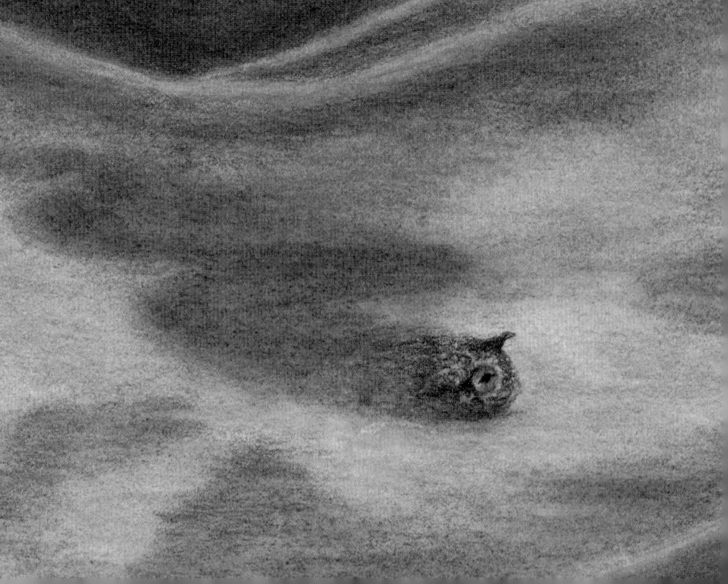

IN A SUBURB OF LONDON

What's going on? The children rush to their windows just in time to see two red foxes that have sent a trash can crashing! Foxes are often the center of controversy in England. Some people consider the clever, agile little mammals a nuisance for raiding garbage and stealing chickens, while other people want them to be protected.

A golden orb-weaver spider is busy rebuilding her beautiful web, with silk thread that comes from her own body. The web was broken during last night's encounter with the foxes. But the British glowworms, which American children call fireflies, aren't disturbed by the foxes. They go on with their ritual mating dance, flashing their lights that give off no heat.

The sweet scent of English primrose has lured several moths to the garden, including a sycamore moth and a brindled beauty. Also on the night-shift is the owlet moth, whose acute hearing helps him avoid the brown long-eared bat patrolling the sky. The British are very protective of their bats; in England it is illegal to move a bat out of its roost.

The owlet moth must also watch out for a scops owl, the smallest of the European owls. Like the long-eared bat, this silent hunter is an *insectivore,* an insect eater.

That hungry hedgehog on the ground is another insectivore. Now he is grunting and snuffling his way toward the earthworms he has sniffed near the dripping faucet. Sharp, short spines on the hedgehog's back help protect him from his enemies, such as the foxes. When in danger, he rolls himself up into a tight, spiny ball. The hedgehog passes by the pink nose of a tiny-eyed mole, busy pushing up from her tunnel crossing the backyard. She, too, would enjoy a dinner of earthworms.

As for tonight's moonlight serenade? The children are treated to bubbly chirps from their own caged house sparrow which has been awakened by the foxes. The tree crickets outside join in the song. Tree crickets are great tricksters. By raising and lowering their wings when they "sing," they can make their enemies think they are somewhere else, far away in the night.

IN THE SONORA DESERT

Now that the air has cooled, it is time for the Sonora Desert's nightshow to begin. The children, inside their adobe home on the outskirts of Tucson, squint out their picture window.

But hidden from their view, an elf owl peeks out of an abandoned woodpecker's hole in a saguaro cactus. Smallest of all owls, he wants to be sure that it's safe to leave his daytime nest. A long-nosed nectar bat moves from blossom to blossom at the saguaro's top, carrying out the important task of dispersing the cactus's pollen, while she gets her meal of flower nectar.

Suddenly, there's a burst of activity. On the ground, a little furry pack rat scuttles back to his home, a rock crevice with bits of cholla cactus piled around it. The pack rat's latest find, a shiny gold locket, is in his mouth. A lanky black-tailed jack-rabbit, her huge ears twitching, bounds back to her hideaway. A squeaking banded gecko, hunted as a crunchy snack by many desert animals, dashes behind a rock. And a shy, handsome ringtail cat, with his bushy striped tail, springs up to the safety of the mesquite tree.

Why such a rush? An alert coyote is surveying the scene from a nearby hill. Coyotes are *omnivores*, animals that eat both animals and plants. In fact, if

this wild member of the dog family can't find any prey, he will fill up on garbage or even leftover dog food. The coyote is one of western America's most adaptable inhabitants.

But a family of coarse-haired javelinas continues happily chomping away on the fruit of the prickly pear cactus. These nearsighted, strong-smelling wild pigs aren't worried about the coyote. He's no match for so many of them. And the large-tusked males can sometimes be vicious.

A poisonous western diamondback rattlesnake slithers away from the noisy javelinas. Two heat-sensitive organs on the snake's face tell her the size and location of other creatures in the dark by the body heat they emit.

On a rock, a fierce-looking hairy spider known as a tarantula searches for beetles. And a bright-eyed Colorado River toad, the largest toad in North America, hops about near a spiny ocotillo plant that has been stripped of its leaves by a colony of nocturnal leaf cutter ants. They, too, are part of the secret life of the Sonora Desert at night.

IN AN INDIAN VILLAGE

On a misty night during India's moonsoon season, everyone in this village near Calcutta has been warned to stay indoors. The tracks of a tiger, the largest member of the cat family, have been spotted nearby. The endangered tiger is native only to temperate and tropical Asia.

The children peer out from their hut. But they can't see the striped cat with porcupine needles stuck in his paw. The Indian porcupine is already waddling away unharmed. Eventually she will grow new quills to replace the ones lost in her victorious tussle with the young tiger.

Not far from the wounded tiger, a regal-looking Indian cobra spreads his hood and hisses, alarmed at the commotion. Throughout India, people both worship and fear this poisonous snake, sacred to their god Vishnu, The Preserver, one of the supreme gods of the Hindu people.

Sensing trouble, a nervous little musk shrew zooms into his burrow. The smallest carnivore in the world, the shrew will even attack mice larger than himself. His excellent hearing, keen sense of smell, and strong musky odor help to protect him against enemies.

A dwarf flying squirrel, one of the few nocturnal squirrels in the world, glides through the trees. And a long, thin-legged panther gecko is in luck! The slow-moving, tailless slender loris that was stalking her has abandoned the chase and moved on. Slender loris are primates, the order of mammals that includes monkeys, apes, and humans. Like monkeys, their feet function as another pair of hands.

Overhead, one of the largest of all bats, the Indian flying fox, flies toward the village. These flying mammals use their large eyes, rather than echolocation, to help them maneuver through the dark. Their favorite food is overripe fruit. By eating such unusable crops, the flying fox keeps away many insect pests otherwise attracted by the rotten fruit.

Here, as in many other Indian villages, mongooses are kept to get rid of rats. And, like the brilliantly colored peacock sleeping in the children's yard, the lithe mongoose also protects against snakes. Should a mongoose battle a cobra, the mongoose will usually win. But the children don't want *their* long-tailed pet fighting any snakes. They hug their mongoose closer, and keep watch for the tiger through the night's veil of mist.

POPCORN

IN THE HEART OF NEW YORK CITY

Many children in Manhattan look out from their apartment house windows as darkness falls over Central Park.

By straining their ears, they can hear the chirps of the crickets and the distinct "Katy did, no she didn't" of the eastern katydid. They can also hear the cooing of city pigeons that roost on ledges outside apartments and perch on benches in the park. A strange bark reaches the children's eager ears. They are willing to bet that it came from the sea lions they saw last week in the Central Park Zoo across the park.

But usually nature's sounds are drowned out by city noises, like the music from "boom box" radios and the rumble of late-night traffic.

By squinting, the children can almost see a big brown bat feeding on the flying beetles known as June bugs that are drawn by the streetlight's glow. A graceful fork-tailed barn swallow, one of many birds that use Central Park as a flyway, also flutters near the lamp.

"Who who who who" shrieks a brick-red screech owl, minding her fuzzy young owlets in a tree. She is giving her shrill "beware" to a terrified tropical macaw. An exotic jungle bird in Manhattan? No doubt the poor thing is an escaped pet.

A tiny field mouse scampers away from the remains of an afternoon picnic under the tree. But two secret residents of the park are not yet done with their food raid tonight. One of them is a fluffy raccoon. The other is a ratlike opossum, North America's only native marsupial. He will roll over and "play dead" if caught in the open. However, most children have no idea that nocturnal animals such as raccoons and opossums dwell well hidden in the very heart of Central Park.

DOWN IN THE EVERGLADES

The Seminole children are returning home from a fishing trip with their father in the Florida Everglades, where the Seminoles as well as the Micosukee Indians live.

Those noises like distant thunder that the children hear actually come from an alligator. These heavy, swampland reptiles make a variety of sounds. Tonight's burpy bellows are made by a new mother. The eggs she has guarded since June are starting to hatch and squeak! Now she must clear away the plant debris covering their nest, so that her hatchlings can reach the water. Alligators lay as many as fifty eggs at a time and will protect their young until they can fend for themselves, which may take as long as a year or more.

The snapping turtle nearby is a less tender mother. As soon as this reptile lays her eggs, she abandons them forever. Now she lies motionless, a piece of wormlike flesh sticking out of her mouth. This is her natural bait to lure small fish, turtles, and water snakes to her for dinner.

The speckled king snake, a large snake with shiny scales, is also after a tasty watersnake, a poisonous pit viper known as the water moccasin. The king snake plans to sneak up on his prey, coil quickly around him, and then squeeze hard until the victim suffocates. If he's successful, it will take the king snake quite some time to swallow his unfortunate meal.

As fireflies flicker in the eerie darkness, a barred owl hoots what sounds like "Who-cooks-for-you-all." And a great blue heron squawks suddenly, disturbed from her rest by an opossum.

From a dwarf cypress, the opossum and her babies peer down at the performers in a lively frog band. The pig frogs grunt, the gopher frogs snore, and the chorus frogs trill. Each has his own sound to call his mate and to warn other frogs away from his territory. Pig frogs have very tasty legs. The children often help catch them to sell to restaurants in the city.

Suddenly, a human-like scream splits the darkness. It is the wail of the panther, called a cougar or mountain lion in other parts of America. One of the panther's chief prey, a white-tailed deer, darts behind a tree, her tail straight up in fright. The human family also quickens its pace. The children can see their new house, with their old, thatched roof "chickee" next to it. They are glad to be almost home at last.

IN A MEADOW IN THE ALPS

It is August already. Near a mountain village in Austria, a roe deer and her twin spotted fawns nibble on the bark of a tree. The smallest of Europe's native deer, they eat facing the wind, so they can quickly pick up the sounds and smells carried by the moving air. Soon the roes' reddish-brown summer coats will turn to a grayish brown for the winter.

Overhead, a beautiful white-faced barn owl carries a tiny fat dormouse back to her owlets nestled in a barn. In Roman days, people raised fat dormice and cooked them as a great delicacy.

On a fence post, cicadas are undergoing a great change. After several years underground, these insects have climbed out to split their ghostly looking outer skins and emerge as winged adults on this very night. In the next few weeks, the cicadas will sing, mate, and lay their eggs, so that a new generation will be on its way before these die in the fall.

The fence is in the path of a nervous little least weasel, his keen nose to the earth, who intently follows a rodent's trail. The least weasel's slender body allows him to squeeze easily into the narrow openings of rodent burrows and capture his favorite prey.

The same fence provides a munchy snack of wasp grubs for a silver gray badger. The badger's thick fur protects her from the wasps' stings. An adaptable mammal that eats a variety of foods, badgers can be found in many places, including North America and Asia. They are excellent housekeepers, regularly cleaning their nests made from dry leaves inside their neat underground dens. Like the least weasel, the badger belongs to the weasel (or mustelidae) family. Ferrets, martens, skunks, and otters are mustelids, too.

The children camped in the meadow pass around a pair of binoculars, hoping to see the red fox. This morning they found his droppings on a stump, left as territory markers, and saw his tracks in the mud. Perhaps the fox has already returned to his den, or "set," with a rabbit or a mouse for his mate and "kits."

Though the children don't see the fox, they can't miss the music in the meadow tonight. Loudest of all is the lively churring of the nightjar. Every camper would love to see this nocturnal songbird, but for now they must be content with only her call, as the first quarter moon of August shines upon them all.

ON THE AFRICAN SAVANNAH

In southern Kenya, a group of Masai boys timidly peek through the thornbush fence, placed around their "manyatta" (or village) to keep out the wild beasts. The strange noises of twilight fill the boys with fright.

Not far away, a lame giraffe had been snacking on the leaves of an acacia tree dotted with weaver-bird nests. Suddenly, two lionesses bound out of the underbrush in hot pursuit of the giraffe. Among lions, the females usually do the hunting. Unseen by the boys, the lionesses' hungry cubs are watching tonight's lesson in co-operative hunting. Once the females have pulled down their prey, the cubs will join in the meal.

The chase has also been noticed by a pack of spotted hyenas, intent on robbing the lions of the kill. But the lions will probably keep the fierce hyenas away until they have had their fill.

Termite mounds as high as trees dot the ground. The tireless termites enrich the savannah's soil by recycling dead matter as they build their homes. But an aardvark is destroying one of their nests to reach the insects inside with his sticky tongue. This strange insectivore with a kangaroo's body, a donkey's ears, and a pig's snout, laps up termites by the hundreds until he feels well fed.

The children shudder as the leopard who had been sleeping on the branch of a tree now awakens with a terrifying yawn. The meat on the antelope carcass, which this large spotted cat has been eating for several days, is gone now. So she, too, must hunt tonight. She may dine on a gazelle, a hyaena, or perhaps that aardvark will do.

A poisonous black mamba, the world's fastest snake, is also in a tree, ready to lunge at unwary prey. The black mamba can chase potential victims at a speed of up to seven miles per hour. A human bitten by a black mamba will surely die unless treated within minutes.

Listen! What made that grunt? A red-gold bushbuck, one of the many African antelopes? And that snort? Could it be from a fleet-footed dik-dik, the smallest of all antelopes? The Masai children have much to learn to survive in the savannah where they hunt and raise their goats and cattle.

IN THE MIDWEST LAKE REGION

Early in September, in northern Minnesota, two children spend the last night of their vacation in their treehouse. Their flashlight's glow reveals two baby raccoons exploring the rowboat while their mother fishes for crayfish. The raccoon family must eat extra food now to store up fat for the cold months ahead. Northern raccoons spend much of the winter sleeping. Once spring comes, the baby raccoons will be off to find their own dens.

Startled by the light, a pair of curious otters interrupt their play. Otters stay active all winter, feeding on fish and crayfish found under the ice. The thick-coated muskrat, returning to his home of reeds, cattails, and mud for the night, will also remain active throughout the year.

Close to the summer cabin, a porcupine nibbles on the handle of an ax. Porcupines love the tasty salt left on the handle by human perspiration, though they normally eat leaves, tree bark, and small animals. Two striped skunks are rummaging through the trash cans again. Not very picky, these omnivores will eat just about anything they find—fruit, cookie crumbs, or chicken bones.

A black bear pokes his huge head out of the thick shelter of the forest. He, too, wants to investigate the trash cans. Though he looks ferocious, his diet is mainly berries, plants, fish, and honey when he can find it. But he won't bother the other visitors here. The porcupine's sharp quills and the skunks' strong, offensive odor keep even burly bears away.

At the children's eye level, a great horned owl perches on a nearby tree. Higher in the sky is a rare and wonderful sight; migrating cliff swallows are silhouetted against the moon as they travel from Canada to their winter home in South America.

With September's chill in the air, all the animals must prepare for winter. While some migrate to warmer climates, others get ready for hibernation, a deep sleep-like state. Most grow especially thick winter coats.

Ice will soon cover the pond. Snow will blanket the cabin and the treehouse. The nights will be cold and silent. Like the raccoons, the bear and the skunks will snooze away much of the winter in a state of partial hibernation. And, as the thermometer drops lower and lower, even the otters, the muskrat, and the porcupine — just like the children themselves — will no longer be seen readily venturing . . . out in the night.

SELECTED READING

Ditmars, Raymond Lee. *Snakes of the World*. New York: The Macmillan Company, 1968 (seventeenth printing).

Duplaix, Nicole and Simon, Noel. *World Guide To Mammals*. New York: Crown Publishers, 1976.

Durrell, Gerald. *The Amateur Naturalist*. New York: Alfred A. Knopf, 1983.

Grzimek, Dr. H. C. Bernhard. *Grzimek's Animal Life Encyclopedia*. New York, Cincinnati, Toronto, London, and Melbourne: Van Nostrand Reinhold Company, 1971.

Hegner, Robert. *Parade of the Animal Kingdom*. New York: The Macmillan Company, 1935.

Jenkins, Marie. *Kangaroos, Opossums and Other Marsupials*. New York: Holiday House, Inc., 1975.

Kondo, Herbert, Editor. *The Illustrated Encyclopedia of the Animal Kingdom*. New York: The Danbury Press, 1968.

Nowak, Ronald M. and Paradiso, John L. *Walker's Mammals of the World*. Baltimore and London: The John Hopkins Press, 1983 (fourth edition).

Library of Congress Cataloging in Publication Data
Liptak, Karen.
Out in the night/by Karen Liptak; illustrated by Sandy Ferguson Fuller.
p. cm.—(Harbinger House juvenile natural history series)
Summary: Introduces the habits and habitat of nocturnal animals in fourteen locations around the world.
ISBN 0-943173-19-1: $15.95. ISBN 0-943173-31-0 (pbk.): $8.95.
1. Nocturnal animals—Juvenile literature. [1. Nocturnal animals.]
I. Fuller, Sandy Ferguson, ill. II. Title. III. Series.
QL775.5.5.L55 1989 591.5—dc 19 89-1833

The author wishes to thank Dr. Ross McPhee, Dr. Brian Robbins, Dr. Donna Howell, and members of the University of Arizona faculty for graciously helping to authenticate this text. And thanks to my editor, Linnea Gentry, whose sons, Timo and Blake, inspired this project.

The illustrator wishes to thank her family, Harbinger House and Maurice Sendak because he is her Mozart.